王劲韬　著
By Wang Jintao

Theme Scenes

主题场景

景观设计手绘教程
Hand Drawing
for Landscape Design

中国建筑工业出版社
CHINA ARCHITECTURE & BUILDING PRESS

前　言

　　本册主题场景记录了作者在长年设计教学与旅行中的些许感悟与感动，从景观感知的角度出发，提出景观与人文地脉、乡村美学等概念结合，分别从城市、公园、滨水绿地、大型园展意象性总图的表达，这类图纸在大型公共项目的表现中占有独特的地位，往往能够反映规划者对项目总体把握的能力，更类似于项目总体印象描述。

　　本书后两章则分别从古典园林表达和农业景观表达两个方面集中体现了景观主题场景表现的艺术可行性，即用设计师、建筑师类似速写渲染的方式如何表现出具有一定感染力的景观表现图纸。题材则选取了大家所熟悉的中国古典花园、欧洲农田景观、大地景观和庄园别墅等，是作者近年来在科研活动的系列速写，是从多角度表现同一主题场景（如一座意大利花园或一座中国花园）的有益尝试。同样希望这些有感之作对读者感悟风景的人文、地域之美提供借鉴。

Foreword

Theme scenes in this volume are a record of the author's thoughts and perceptions in teaching and traveling in the past years. From the perspective of landscape perception, they suggest a combination of landscape and concepts like cultures, geographical layouts and rural aesthetics, respectively by way of expression in general imagery plans of cities, parks, waterfront greenbelts and large park exhibitions. Such drawings are of unique importance in the presentation in large public projects, often reflecting the planner's overall understanding of the project, and are more like descriptions of the general impression on the project.

The last two chapters of the book, respectively through the expression of classical gardens and of agricultural landscapes, demonstrate that landscape theme scenes is a feasible way of artistic expression. Quite infectious landscape expression drawings can be presented by designers and architects in a way like sketching and rendering. Themes familiar to the public are selected, including Chinese classical gardens, European farmland landscapes, earthscape and plantation villas. They are a set of sketches of the author's scientific research activities in recent years, as well as a useful attempt to express the same theme scene from different aspects (an Italian garden, a Chinese garden, for example). I hope these results of thoughts will provide references to readers for feeling the cultural and geographical beauty of landscapes.

目 录

008　　第 1 章　　城市意象及设计的表达

034　　第 2 章　　公园的表达

050　　第 3 章　　城市滨水绿地的表达

072　　第 4 章　　大型展园的表达

090　　第 5 章　　古典园林的表达

108　　第 6 章　　农业景观的表达

Contents

008   Chapter I   Expression of City Imageries and Designs

034   Chapter II   Expression of Parks

050   Chapter III   Expression of City Waterfront Greenbelts

072   Chapter IV   Expression of Large Exhibition Parks

090   Chapter V   Expression of Classical Gardens

108   Chapter VI   Expression of Agricultural Landscapes

第1章
城市意象及设计的表达

Chapter I
Expression of City Imageries
and Designs

纽约

纽约市内高楼林立,帝国大厦、时代广场、自由女神像、中央公园等是纽约的标志。纽约意象的景观表现应重点在对其高楼、窄街的描绘。由于高楼矗立,光线几乎无法照进窄街内,街道上的光影由内而外渐浅,高楼上的光线由上而下渐深,从而进一步加强了窄街的效果,加之视线由低到高的收缩,使得高度围合的容积空间感得到加强,从而营造出了一种高耸、深邃的感觉。

本页选择了作为世界金融中心的纽约最具典型性的画面——华尔街景观。分别由外向内看华尔街中心区,由内向外看证券大厅,以及从联邦大厅华盛顿雕像方向看华尔街,刻意表现了阳光射入金融中心的经典场景。建筑上的星条旗能点出地域特征,而宝贵的一线天使高楼环抱的街道不那么死气沉沉。

New York City

    The New York City is a forest of skyscrapers, with the Empire State Building, Times Square, the Statue of Liberty and Central Park as its landmarks. The landscape expression of NYC's imagery shall focus on the description of the high-rises and narrow streets. Kept out by the high buildings, light can barely shine through these streets, resulting in an effect where shadows become shallower and shallower from the inside to the outside on the streets, but deeper and deeper from top to down on the buildings, making the streets seem even narrower. The shrinkage of eyesight from low to high further strengthens the sense of highly-enclosed spaces, creating a feeling of towering and depth.

    On this page the landscape of Wall Street is selected as a typical picture in NYC, the world financial center. Scenes of different angles are presented, namely from outside to inside to see the center or Wall Street, from inside to outside to see the Securities Hall, and from Washington statue at gate of the Federal Hall to Wall Street. The classical scene of sunlight bursting through the Financial Centre is highlighted. The stars and stripes flag on the building indicates the geographical character, and the precious thread of sky gives the street surrounded by skyscrapers some vitality.

光影模式：① 由内而外渐浅，光
② 光度由下而上渐浅，
③ 视线由低到高渐收

两种感觉：高耸 → 深远

城市街景欣赏——纽约时代广场
View of the City street — Times Square, NYC

纽约时代广场
Times Square, NYC

纽约时代广场
Times Square, NYC

波士顿

波士顿位于美国东北部大西洋沿岸，是美国最古老、最有文化价值的城市之一，它曾经是一个重要的航运港口和制造中心，如今是高等教育和医疗保健中心。在景观上，波士顿的港区、查尔斯河非常具有地方特色。以下几张图展示了波士顿港口码头与大开挖线状公园带、波士顿公地与中心区关系以及查尔斯河和港口城市建筑群，通过鸟瞰图来解读城市能更加明确形象化的空间意象特征，绿带的分布、河流的走向、城市的天际线等要素，可以通过人为色彩，即主观的处理，而得到更清晰的分区。

波士顿长码头滨海区域鸟瞰
Bird's eye view of coast area in the Long Wharf, Bost

波士顿翡翠项链"第一环"——波士顿公地公园鸟瞰
Bird's eye view of Boston Commons, the "first link" of Boston's Emerald Necklace

Boston

Located at the Atlantic coast in the Northeast United States, Boston is one of the nation's oldest cities with the highest cultural values. It was once a major shipping port and manufacturing center, and now the center of higher education and health care. From the perspective of landscape in Boston, we can find distinct local characteristics in the port area and the Charles River. The following pictures show the Boston's ports and docks, linear zone of parks made through heavy excavation, relations between Boston Commons and the central area, and the Charles River and port city building complex. When the city is shown in a bird's eye view the visualized characteristics of space imageries become clearer. The distribution of greenbelts, running of river courses, city skyline and other elements are partitioned in a more distinct way through artificial coloring, which is a subjective treatment.

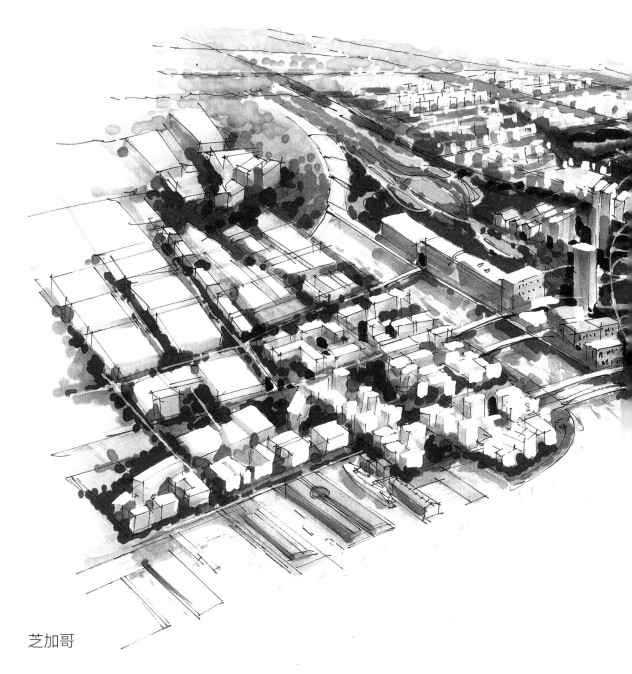

芝加哥

　　芝加哥地处北美大陆的中心地带，是美国最为重要的铁路、航空枢纽，同时也是最重要的金融、文化、制造业、期货和商品交易中心之一。白金汉喷泉位于芝加哥格兰特公园内，是世界上第一大照明喷泉，其中央水柱喷射高程可达四五十米。图中对于喷泉的表现采用了衬托的方法，前方用深蓝褐色处理的道路、中部墨绿色处理的树林以及后方棕灰色系处理的建筑和天空都是为衬托喷泉的水柱埋下的伏笔，喷泉水的表现除局部的留白外，大部分是在画好配景后用修正液和油漆笔对水花进行的再勾勒，从而将如烟花般的水花生动地表现出来。

　　芝加哥海军码头位于芝加哥的密西根湖边，其北侧是芝加哥河，沿河边一直走可到人工岛的尽头，站在尽头广场上向湖上望可以看到两道防波堤远远伸向湖中，围成一个港湾，两座灯塔遥相呼应。码头南侧有许多豪华游艇，另一侧是一长排错落有致的建筑，此外还建有摩天轮和旋转木马等游乐设施。对芝加哥的手绘表现图中多围绕海军码头展开，有从湖面方向鸟瞰码头全景的，有从阿德勒天文馆方向望向码头的，也有海军码头回望身后建筑的，还有从建筑方向俯视码头的……而在这些画面中也都未离开西尔斯大厦、格兰特公园的身影。

芝加哥河入口处城市总体鸟瞰
Panoramic aerial view of Chicago from the entrance of Chicago River

Chicago

Located in the heart of the North American continent, Chicago is the most important railways and aviation hub in America, as well as one of the most important financial, cultural, manufacturing, futures and commodities trading centers. Buckingham fountain lies in Chicago's Grant Park. It is the world's largest illuminated fountain, with the central water spraying up to forty to fifty meters. In the drawing the fountain is expressed with the help of contrasting background. The dark blue and brown road in the front, dark green woods in the middle and brownish grey buildings and sky in the rear, all are the background prepared to set off the sprays of the fountain. Only a small part of the fountain water was left blank, and the rest part the water splashes was re-outlined with white-out and paint marker after the background is drawn, so that the splashes come out vividly just like fireworks.

Chicago's Navy Pier is at the edge of Lake Michigan, with the Chicago River to the North. Walking along the river you can come to the end of the artificial island. Standing on the square there and looking toward the lake you will see two breakwaters reaching far into the lake, forming a harbor, with two lighthouses echoing each other at a distance. At the south of the pier there are a number of luxury yachts, and on the other side there is a long row of buildings in picturesque disorder, together with entertainment facilities including a ferris wheel and merry-go-round. Navy Pier is the main theme of the hand drawings of Chicago. Some are panorama overlooking the pier from the lake, some sketches watching the pier from the Adler Planetarium, some looking back from the pier toward buildings behind, and some vertical view from the buildings toward the pier. In any of these drawings you can see Sears Tower and Grant Park.

芝加哥伯纳姆公园滨水区域总体鸟瞰
Panoramic aerial view of the waterfront zone in Burnham Park, Chicago

芝加哥河东部公园绿地总体鸟瞰
Panoramic aerial view of the greenbelts of parks in the east of Chicago

芝加哥海军码头综合娱乐区鸟瞰
Aerial view of the comprehensive entertainment area of Navy Pier, Chicago

巴尔的摩内港

巴尔的摩是美国大西洋沿岸重要的海港城市，独立战争期间曾一度是美国战时首都，故具有丰富的历史遗迹，有"不朽城"之称。图中手绘表现的是巴尔的摩港的一隅，画面近景的矩形水面与远方淡蓝的天空融为一体，突出了水对城市的影响，水边成列树林的墨绿与远方建筑上的几抹亮黄又赋予画面以生机和活力。画面色彩由中部向外围逐渐减淡，增加了场景的延伸感，使人由一角而联想到整个城市。

Baltimore Inner Harbor

Baltimore is an important port city on the Atlantic coast in the United States, and was once the wartime capital of the States during the War of Independence. It has a lot of historical heritages and is known as the "Eternal City". The hand drawing expresses a corner in Baltimore. A rectangular water surface in the foreground fuses into the blue sky far-off, highlighting the effects of water on the city. Rows of dark green trees by the water and a few touches of bright yellow above the building in the distance endow the picture with vitality and vigor. Colors on whole picture becomes lighter and lighter gradually from the center to the outer edges, bringing about the feeling of extension, so that anyone seeing this corner will think of the whole city.

浙江衢州信安湖新城规划总体鸟瞰
Panoramic aerial view for new city planning of Xin'an Lake, Quzhou, Zhejiang

  建筑为主的城市景观在园林景观设计中独立使用得较少，其主要功能是作为大规模公园绿地的背景，从而表达出城市与绿地相互渗透的概念。手绘表现中建筑多以组群形式出现，前景有大片绿色或临水开阔空间，与正式意义上的建筑表现相比，这种组群建筑表现一般比较简洁，省略了建筑图中必需的窗、幕墙等细节，只留下大的形体和色彩意象，虽然寥寥几笔，但用笔十分肯定，在反映城市意象的特大场景中，建筑立面上往往只保留两种基本元素和色块，一是明暗，一是投影，而投影需要特别注意。位于近处的建筑立面上的投影应敢于用深、重的色块，笔触要实，以突出建筑在特定的城市空间中的落位和层次感。

  此外，还要注意楼群之间色彩的渐变要协调。由近景的暖色到远景的冷灰，由底楼的深棕色到楼顶的亮黄色，此类的单色渐变、多色渐变以及用马克笔和彩铅交替变化退晕组合的方式非常多，可视图面复杂程度以及场景色彩要求选择使用。如在以城市意象为主要表达的图纸上，可选用多种色彩混合渐变，最后以黑色或深蓝（近于无彩色）画出强投影，归拢所有的色彩。而在表达绿地为主的城市背景中，一般仅仅选用 1~2 种灰调作近于素描效果的单色渐变，最后将天空和楼群一并着色，也就是把环境色作为建筑的统一意象色彩，起到背景作用即可。

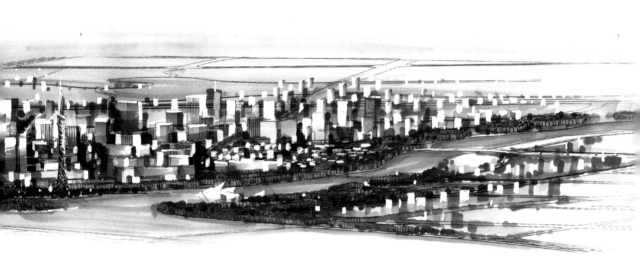

Cityscape dominated by buildings is seldom independently adopted in landscape architecture. Instead, it is mainly used as the background of vast park greenbelts to express the mutual penetration of urban areas and green spaces. In hand drawing, buildings are presented mostly in groups, with large green space or open space nearing water as the foreground. Compared with buildings in the formal sense, such building groups, or building complexes, are often expressed in a more concise way. Windows, curtain walls and other details necessary in building plans are omitted, with only the general shape and color imagery rendered, through just a few yet assured strokes. In those extra-large scenes reflecting the city imagery, generally only two basic elements and color blocks are remained on the building facades, namely, light and shade, and projection. Special notice: be bold to use dark and heavy colors in depicting the projection on the facades of nearby buildings, and the strokes shall be solid in order to emphasize the position of building in a particular urban space and to express layering.

Additional attentions shall be paid to coordinate the color gradient among buildings. From warm colors at close range to cold grey at long range, from dark brown on the ground floor to bright yellow on the rooftop, there are a lot of one-color and multi-color gradual changes, as well as the combination use of marker pens and color pencils to produce the effect of alternating changes of colors. Just choose those suitable for your picture complexity and the scene's requirements on colors. For example, when the drawing is mainly to express the city imagery, you can choose mixed gradual change of multiple colors, and use black or dark blue (almost colorless) to draw up a strong projection in the end to put together all the colors. While in a background aiming to express the green space, generally only one or two achromatic colors will be adopted, so as to realize one-color gradual change that is closed to the effect of a sketch; finally the sky will be colored along with the building complex, that is, the environment color will be used as a unified imagery color for the buildings so that they act as a background merely.

浙江衢州信安湖新城中央核心区城市之门鸟瞰
Aerial view of the City Gate in the core area of the new city of Xin'an Lake, Quzhou, Zhejiang

深圳高铁站区域绿谷规划鸟瞰
Aerial view of the plan of the Green Valley in the high speed rail station area, Shenzhen

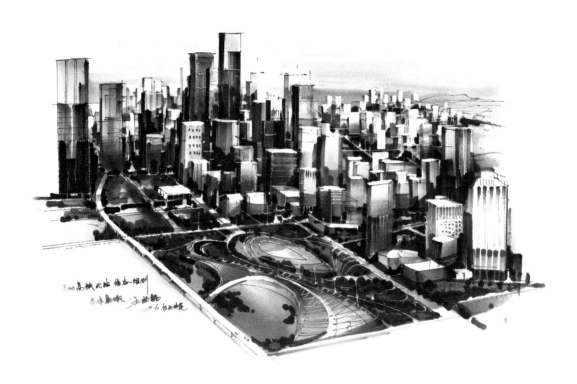

深圳高铁站区域绿谷规划鸟瞰
Aerial view of the plan of the Green Valley in the high speed rail station area, Shenzhen

市民的客厅——同在一片屋檐下

此类景观项目大多涉及政府形象和市民使用等诸方面的问题,在此方面体现出极大的灵活性和模糊性。

本设计为秦皇岛金梦海湾的市民广场,及城市形象区景观意象,考虑到使用者对场地的认知和应用,对同一场地表现出春夏秋冬四季的变化,同时考虑到白天和夜晚使用的不同的景观效果。

Citizens' Living Room — Under the Same Roof

Most of such landscape projects involve various issues like the government's image and usage by the public, and therefore they feature great flexibility and ambiguity.

Here we take the design of the Civic Square at Golden Dream Bay in Qinhuangdao and the landscape imagery of the city image area as the example. Taking into account the users' understanding and application of the space, we focus on the expression of changes on the same site through the four seasons, while keeping an eye on producing different visual effects for daytime and nights.

该类场景中建筑群的表现过程是：先通过简易线图画出对天际线最具影响力的建筑群及城市前景，后部大量楼群是后期"点"出来的；然后进行第一遍着色以完成落影、分层的工作，影子是建筑物"落地"的重要步骤，阳光自城市上方射入，上浅下深的色彩布局自始至终都要保持好；而后要加画中间层（即固有色层），前景为亮黄色，后景用灰蓝色，在色彩深浅、饱和度两方面对图像进行了主观的人工处理；最后加画天空和近景的水面及船只，完善场景的层次，点出远景的建筑群（该过程宁简勿繁，点到为止，忌处处闪光，杂乱无章）。

Here is how the building complexes in such scenes are expressed: firstly, simple lines are used to draw up the building complexes and city foreground that have the most influences on skyline, and the large number of buildings in the back are to be drawn through "dotting" later. Then the coloring to complete shadowing and layering is to be done first. Shadow is a significant part for buildings to "fall on the ground". Sunlight shoots down from above the city, so it is important to always keep the colors in a "light on the top and dark at the bottom" pattern. The next step is to paint the middle layer (proper color layer), with bright yellow as the foreground and grey-blue the background. The shade of colors and degree of saturation is adjusted subjectively. Finally, the sky as well as water and vessels in the foreground is painted to complete the graduation in the scene, and the building complex at long range is added through "dotting" (which should be better simple than complicated. Just play touch to avoid too many focuses and disorder).

第 2 章
公园的表达
Chapter II
Expression of Parks

公园作为城市的主要公共开放空间在满足城市居民休息、游览、锻炼、交往和举办文化活动的同时还起着改善生态和预防灾害的作用。作为城市中的绿地，城市公园周围多被建筑包围，其内部有较明确的功能分区，空间组成上有满足集会需求的开敞空间，也有满足休憩需要的私密空间，而植物配置上除作为公园骨干树种的高大乔木外，也会相对较多地种植花、果、叶具有观赏价值的小乔木或灌木类树种以满足人们的观赏需求。

Parks are the main public open spaces in cities. They not only provide the places for the people to rest, visit around, do exercises, communicate with each other and hold cultural events, but also play a role in improving the ecological environment and prevent disasters. As green spaces in the city, urban parks are mostly surrounded by buildings. Spaces inside them are clearly divided for different functions. There will be open spaces for assembling activities, and private spaces for resting. As for planting arrangement, except that tall trees are planted as the main species, small trees or shrubs with ornamental flowers, fruits and leaves are also grown to meet visual needs of the visitors.

植物的季相变化表达
Expressing seasonal changes of plants

植物的季相变化表达
Expressing seasonal changes of plants

植物的光与影表达

Expressing light and shadows of plants

植物的主题性表达
Expressing subjectivity of plants

039

深圳平湖自行车公园绿道规划示意图
Greenway planning schematic map of Pinghu Bicycle Park, Shenzhen

北京西山景观规划从新中国成立以来就持续进行，其核心是围绕着现有的皇家园林古迹进行植物的总体品种规划，突出了西山的春秋两季的季相特征，体现出具有标志性的植物景观：春季以山杏、山桃为主，形成林木蓊郁，云蒸霞蔚的效果；秋季则以秋色叶为主，黄栌和槭树类植物把西山点缀得火红一片，本设计依据佛牙舍利塔古迹表现了西山八大处上下三台的春季景观效果。

Beijing Western Mountains landscape planning program was launched at the founding of PRC, with the core task to make the overall plan of plant varieties based on existing ancient royal garden relics. The plan is to highlight the seasonal changing characteristics between the spring and fall of the Western Mountains, present the symbolic vegetative landscape, which is dominated by apricot and mountain peach in the spring, forming luxuriant green woods within colorful and flourishing clouds, and in the fall there are mainly autumn-leaf trees like smoke trees and maple trees, dyeing the mountains in a world of flames. The design, based on the ancient Buddha's Tooth Stupa, expresses the visual effects of three tiers from upper to lower in Badachu, Western Mountains in the spring.

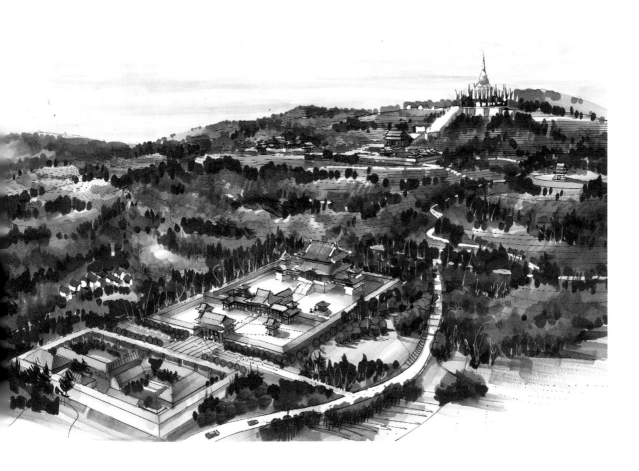

北京八大处国际文化区 第三层台鸟瞰——清凉寺
Purity Temple, aerial view from the 3rd tier of the International Culture Zone, Badachu, Beijing

不同品种的植物往往能够反映出强烈的地域特色和文化特色，作者选择了北方地区特色的松和柏为例，阐释植物对场景地域性特色的影响性。

北方的落叶松反映了中国北部山区的特色，尤其是秋冬景观的意象；古柏的虬枝很好地体现出中国皇家园林厚重的历史文脉与富丽堂皇的气息，对古典建筑环境的烘托起到了决定性作用。

Different varieties of plants often reflect distinct regional and cultural characteristics. The author chose the pine and cypress which are typical in Northern China as the example to demonstrate the influence of plants on the regional characteristics of scenes.

The larch of the northern areas is a reflection of characteristics of mountains in the North China, especially of autumn and winter landscape imageries; curled branches of the ancient cypress perfectly presents the profound historical context and magnificence of Chinese royal gardens, playing a decisive role in highlighting the atmosphere of classical architectures.

植物的地域性表达

Regional expression in plants

这是整个胜溪湖周边景观规划的一部分，景观设计的主要目标在于完善和重塑寺庙周边的山林沟谷景观。本图刻意表现了寺庙的落位、竖向以及周边通过植物设计以后形成的丰富季相色彩，是植物主题性表达的一次较好尝试。

As part of the landscape planning for the area surrounding Shengxi Lake, this design mainly aims to improve and reconstruct the mountain forest and ravine landscape surrounding the temple. This drawing intends to express the temple's location, vertical landscape and surrounding landscape of rich seasonal color changings formed through the vegetation design. It is a successful attempt of thematic expression with vegetation.

山东章丘秀江河滨水公园鸟瞰
Aerial view of Xiujianghe Waterfront Park, Zhangqiu, Shandong

河北曹妃甸湿地北湖区整体鸟瞰
Panoramic aerial view of the north lakeland in Caofeidian wetland, Hebei

第 3 章
城市滨水绿地的表达
Chapter III
Expression of City Waterfront Greenbelts

滨水景观作为城市不可多得的资源和风景,决定着城市景观的特点,同时影响着城市景观的发展方向。凯文·林奇在《城市意象》中提出城市空间的五要素:道路、边缘、节点、地标、广场。滨水公园既是城市的节点,其本身又同时覆盖了这些景观要素,其规划设计应满足塑造水域空间的开放性、增强水域景观的可达性与亲水性、打造交通活动的连续性以及彰显水域景观的文化性四方面的要求。

对于滨水公园的手绘表现,也应重点突出其水与绿地相结合的特点,无论是鸟瞰图还是节点效果图,均不能忽视水的存在以及围绕水所展开的活动。在双龙河滨水绿道鸟瞰图中,主要是要表现出河流的走向与绿地的关系,故手绘时只需用简单线条勾勒出河流、堤岛、树丛的基本轮廓,然后用马克笔确定各要素间的色彩及明暗关系,最后点出场地中的建筑并画出远处的天空,天与水的色彩要相呼应,以营造出水天交接之感。而对绿道局部效果的描绘则要相对细致,近处的行道树、步行路以及路上的游人都需有较为详细的表现,画面远处的河道、树林及天空用相对抽象的色块表现即可。

河北秦皇岛南戴河森林公园总体鸟瞰
Aerial view of Nandaihe Forest Pink in Qinhuangdao, Hebei

Waterfront landscape, as rare resources and sceneries in cites, determines the characteristics of the urban landscape, and influences the direction to develop the latter. In his book " The Image of the City ", Kevin Lunch makes five elements for the city space: paths, edges, districts, nodes and landmarks. Waterfront parks are nodes of the city, and covered the landscape elements at the same time. Their planning and design should be able to meet four requirements. namely building open water spaces, enhancing the accessibility of water scapes and their closeness to water bodies, keeping traffic continuity, and showing the culture of water scapes.

Hand drawing of waterfront parks should highlight the combination of water and green spaces. Whether in an aerial view or a node design sketch, water should not be overlooked, nor shall the activities carried out around the water.

山西吕梁行政区湿地公园
Wetland Park in Lvliang administrative region, Shanxi

上海金山区生活水岸规划——游艇码头区鸟瞰
Residential water-front planning of Jinshan District, Shanghai — aerial view of the yacht dockland

对城市公园全景的手绘表现上应注意对公园周围的建筑及天际线的处理，用公园周边林立的高楼烘托出城市的氛围，从而指明公园所处的区域特征。周边建筑的表现不用过于细致，只需用直线勾勒出建筑物的主体轮廓，再用马克笔画出建筑间的强投影和建筑的明暗面即可。对城市公园局部表现可选取较具特色的建筑小品或林下景观，局部表现应注意近景树的处理，和人物配景的点缀，人物的点缀在反映景观尺度的同时能使画面更富生活气息。

In hand drawings that express urban park panoramas, attentions should be paid to the expression of buildings surrounding the park and the skyline, with high-rises around the park creating the urban atmosphere, which can indicate the regional characteristics of the park. Surrounding buildings need no detailed expression; just use simple lines to depict their outlines, and then draw up the strong projections among buildings and the light and shade on them with marker pens. Local expression of urban parks can be realized through distinctive architectural pieces or forest landscapes. In such local expression attentions shall be paid to the presentation method of trees at close range and people as the background, with the latter reflecting the landscape scale and endowing the picture with vitality.

山东潍坊白浪河滨水公园总体鸟瞰

Panoramic aerial view of Bailang River Waterfront Park, Weifang, Shandong

河北曹妃甸入海口湿地鸟瞰
Aerial view of Caofeidian estuary wetland, Hebei

河北曹妃甸中央湿地鸟瞰
Aerial view of Caofeidian central wetland, Hebei

河北曹妃甸虾池鱼塘湿地鸟瞰
Aerial view of Caofeidian shrimp pond and fishpond wetland, Hebei

河北曹妃甸北湖区休闲度假区鸟瞰
Aerial view of the recreational zone in the north lakeland

河北曹妃甸曹妃湖港区
Dockland of Caofei Lake in Caofeidian, Hebei

河北曹妃甸曹妃湖港区
Dockland of Caofei Lake in Caofeidian, Hebei

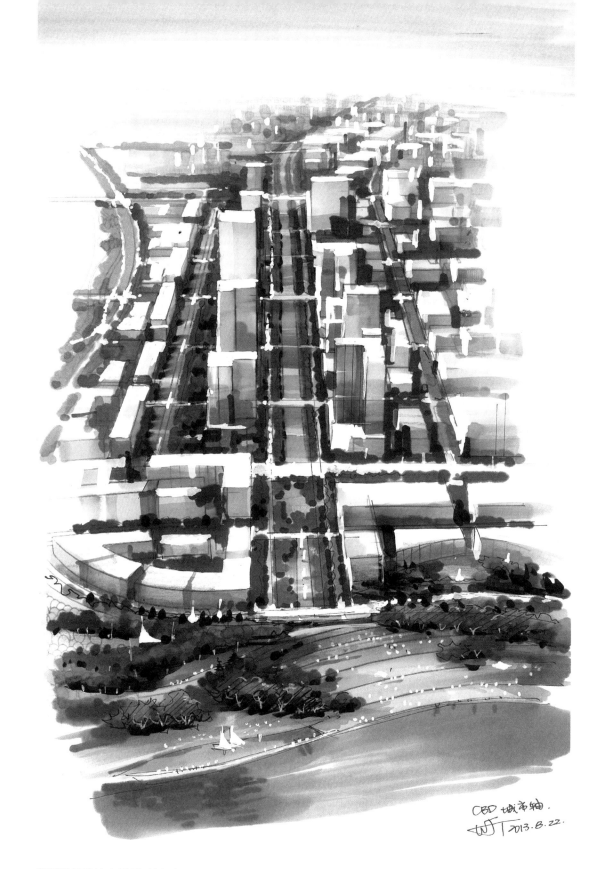

衢州新城规划中央景观绿轴鸟瞰

Aerial view of the central green landscape axis for new city planning, Quzhou, Zhejiang

山西吕梁新城规划行政区鸟瞰
Aerial view of administrative regions for new city planning, Lvliang, Shanxi

山西吕梁新城规划行政区中央公园鸟瞰
Aerial view of the Central Park in the administrative regions for new city planning, Lvliang, Shanxi

上海金山生活水岸滨水休闲区鸟瞰
Aerial view of the residential water-front recreational area of Jinshan District, Shanghai

鄂尔多斯伊金霍洛旗大南沟水库市民公园中央草坪
Central lawn in citizens' park the beside Danangou reservoir, Ejin Horo Banner, Ordos

衢州沿新安江新城规划中央区鸟瞰
Aerial view of the central zone for new city planning along the Xin'an River, Quzhou

阜阳水系规划景观鸟瞰

Aerial view of river system planning landscapes of Fuyang

北京延庆城市水系规划鸟瞰
Aerial view of city water system in Yanqing, Beijing

北京延庆城市水系规划鸟瞰
Aerial view of city water system in Yanqing, Beijing

北京延庆城市水系规划鸟瞰
Aerial view of city water system in Yanqing, Beijing

福建泉州洛阳江水系规划鸟瞰

Aerial view of Luoyangjiang water system in Quanzhou, Fujian

第 4 章
大型展园的表达
Chapter IV
Expression of Large
Exhibition Parks

唐山 2016 世界园艺博览会入口总体鸟瞰
Panoramic aerial view of the entrance to International
Horticultural Expo 2016, Tangshan

唐山 2016 世界园艺博览会总体平面图
General plan of International Horticultural Expo 2016, Tangshan

唐山 2016 世界园艺博览会杏花春雨鸟瞰
Aerial view of "Apricot Flowers, Spring Rain" scene of International Horticultural Expo 2016, Tangshan

唐山 2016 世界园艺博览会孟兆祯先生改造的垃圾山鸟瞰
Aerial view of the "Garbage Mountain" transformed by Mr. Meng Zhaozhen of International Horticultural Expo 2016, Tangshan

唐山 2016 世界园艺博览会主场馆小南湖鸟瞰
Aerial view of the main venue, Small Nanhu Lake of International Horticultural Expo 2016, Tangshan

唐山2016世界园艺博览会孟兆祯先生设计的小南湖煤矸石山观光塔鸟瞰
Aerial view of the sightseeing tower of coal refuse mountain designed by Mr. Meng Zhaozhen in Small Nanhu Lake of International Horticultural Expo 2016, Tangshan

唐山 2016 世界园艺博览会孟兆祯先生设计的堤岛总体鸟瞰局部

Partial aerial view of the "dam island" designed by Mr. Meng Zhaozhen of International Horticultural Expo 2016, Tangshan

唐山2016世界园艺博览会孟兆祯先生设计的堤岛总体鸟瞰局部
Partial aerial view of the "dam island" designed by Mr. Meng Zhaozhen of International Horticultural Expo 2016, Tangshan

唐山 2016 世界园艺博览会大南湖小组树木园鸟瞰
Aerial view of the arboretum at Large Nanhu Lake on International Horticultural Expo 2016, Tangshan

第十二届中国国际园林博览会（长沙）总体鸟瞰
Panoramic aerial view of The 12th China International Garden Expo (Changsha)

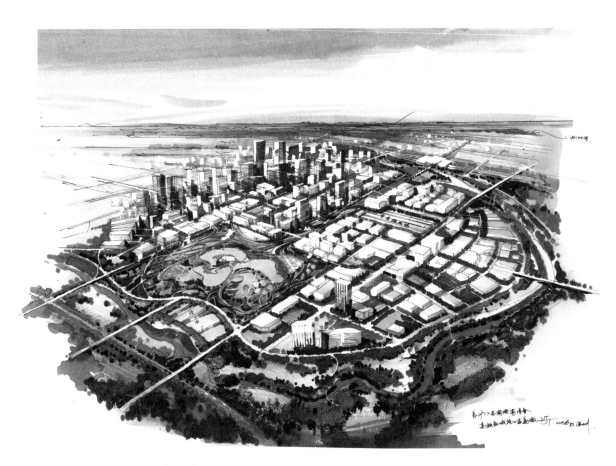

第十二届中国国际园林博览会（长沙）总体鸟瞰
Panoramic aerial view of The 12th China International Garden Expo (Changsha)

第十一届中国国际园林博览会（郑州）——从南水北调干渠方向鸟瞰
The 11th China International Garden Expo (Zhengzhou) —— aerial view from the trunk canal for South-to-North water diversion

第十一届中国国际园林博览会（郑州）——景观塔、轩辕阁鸟瞰
The 11th China International Garden Expo (Zhengzhou) —— aerial view of Xuanyuan Pavilion, the landscape tower

第十一届中国国际园林博览会（郑州）——西北向鸟瞰
The 11th China International Garden Expo (Zhengzhou) —— aerial view from the northwest

第十一届中国国际园林博览会（郑州）——从航空港新城方向鸟瞰
The 11th China International Garden Expo (Zhengzhou) —— aerial view from the airport new city

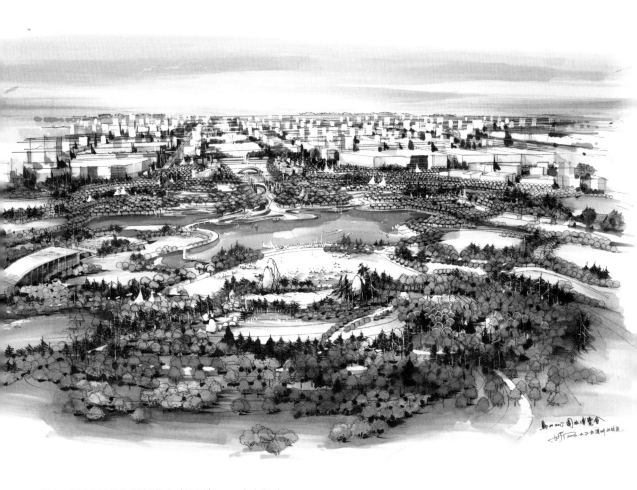

第十一届中国国际园林博览会（郑州）——主山鸟瞰
The 11th China International Garden Expo (Zhengzhou) —— aerial view from the main hill

第 5 章
古典园林的表达
Chapter V
Expression of Classical Gardens

　　北京某四合院的景观改造设计图纸四张，运用了北方古典民居的典型构建和庭院要素进行了重新组合，体现了"天棚、鱼缸、石榴树，先生、肥狗、胖丫头"的人景交流的温馨画面。

　　其中的景观元素主要突出了中国假山石特质和造型树、色叶树的陪衬，建筑上强化了砖雕照壁和小巧的垂花门入口，使甬路、水塘的平面划分丰富了传统庭园的意象，这是作者设计中很少使用水彩渲染的案例，偶一为之别有趣味。

　　Here are 4 landscape renovation design drawings for a Siheyuan in Beijing. The typical structure and courtyard elements of classical residence in North China are re-combined. It shows the warm picture human and landscape: eaves, fish tank, pomegranate trees, sir, fat dog and fat servant girl.

　　Among all the landscape elements, the featured Chinese rockery design, as well as style trees and color-leaved trees as the background, are highlighted; in respect of architecture, brick carved screen wall and the small yet exquisite flower—decorated door entrance become the focus. The planar subdivision realized by paved pathways and pond enriches the imagery of the traditional courtyard. In this example water color rendering is used, which is seldom for the author, for which reason it becomes especially interesting.

北京某四合院景观改造
Landscape renovation of a Siheyuan, Beijing

北京某四合院景观改造
Landscape renovation of a Siheyuan, Beijing

北京某四合院景观改造
Landscape renovation of a Siheyuan, Beijing

古典园林庭院
Classical garden courtyard

093

古典园林花园围墙细节

Details of enclosing walls of the classical garden

古典园林建筑细节
Architectural details of the classical garden

古典园林元素细节
Details of classical garden elements

罗马古迹区的总体鸟瞰

位于卡比多利奥山的大平台上是鸟瞰罗马古迹区的最佳视点，从右侧艾曼纽纪念碑开始，视线经帝国大道，跨越台伯河直达 20 公里以外的弗拉斯卡蒂的连绵山丘。中景处则以巨大的帕拉丁山别墅、山下废墟为主体，视线从远景的大竞技场、提图斯凯旋门与近景的维斯塔小神庙、马森缇与巴西利卡、凯撒艾米莉大堂、农神撒旦庙到达塞维鲁凯旋门，画面近景重点描绘了罗马特色伞松的优美姿态，通过单色马克笔在光面草图纸上的多层叠加，将简单速写材料变化的丰富性大大提高，增加了画面的艺术表现力和历史感。

Bird's Eye View of Roman Ruins

The best bird's eye view of Roman ruins can be obtained from the big platform on Campidoglio, reaching as far as rolling hills at Frascati 20 kilometers way, starting from the Victor Emmanuel Monument on the right side, by way of the Empire Avenue and the Tiber River. With medium view highlighting the huge Palatine villa and the ruins beneath the hill, our eyes are feasted from the Coliseum, Arch of Titus afar and Vestae aedes, Basilica of Maxentius and Basilica Aemilia, Temple of Saturn to Arch of Septimius Severus nearby. The near view gives priority to graceful gesture of featured umbrella pine in Rome, and applies multiple use of solid-colored marker on glossy sketching paper, which substantially enriches variation of originally simple sketch material and adds artistic expression and sense of history to the picture.

梯沃利的艾斯特花园

  作者用了 10 余张小图连续绘制了艾斯特这座文艺复兴名园在栽植、建筑细节和环境方面的特色。这座文艺复兴鼎盛时期的花园曾经以规则修剪和浓郁的古风雕塑闻名于世，但随着权势熏天的依波利托主教的离去，这座伟大的艾斯特家族花园一度无人问津，被荒废达两个世纪之久，再加上欧洲各国古董家、冒险家蜂拥而至地购买与掠夺，这座花园所有珍贵的收藏、雕塑被洗劫一空，而图中的古柏恰恰因为这种无人问津而一度疯长，使一座本以规则著称的花园变得超级自然，园中古柏甚至高达 30 米以上，为意大利全国少有，其姿态之奇崛多变和雄壮也唯有北京太庙古柏才可与之媲美。今日图中所绘，与苍古名园的设计相比成败不足一也。

Villa D'Este, Tivoli

The author depicts planting and architectural details and environmental features of Villa D'Este, a celebrated garden during the heyday of Renaissance Period, with more than ten sketches. Once known for its regular pruning and sculptures full of ancient style, the garden has been neglected for two centuries long, accompanying the departure of the powerful bishop Ippolito. Worse, it has lost all its precious collections and sculptures by those flocked European antiquaries' and adventurers' purchase and plunder. Ironically, cypresses in sketch have unrestrainedly bloomed without due attention, making the originally regular garden super natural. Some cypresses are even up to over 30 meters, which are rarely seen in Italy; some are so distinctly rugged and peculiar that only those in the Imperial Ancestral Temple of Beijing can be comparable. Compared with the garden design, what the sketches depict is only the tip of the iceberg.

意大利园林中的柏树与中国古典园林的有明显不同的特征，中国园林的柏树以纵向出枝为主，老干蟠曲，色彩偏暖；而意大利柏树则反之，分支点极低，横向发展，往往呈大面积的团块状，冬夏季色彩变化不明显，皆呈现葱绿的色彩。

Cypresses in Italian gardens are largely different from those in Chinese classical gardens. The latter ones mainly have vertical branches and warm colors, with old trunks coiling up, while in Italian gardens, on the contrary, cypress trees have much lower branch points, and thus branches often grow horizontally into blocks clumps, and they keep being green from winter to summer, with less seasonal changes in color.

第 6 章
农业景观的表达
Chapter VI
Expression of Agricultural
Landscapes

  本章以人与自然共同影响下的景观为例，表现一组以农业景观第二自然为主导的风景，同时这也是"诗意栖居"最为真实和历史最为悠久的载体。这样的创作活动源于 2008 年作者参与的以色列耶路撒冷的规划。当年的一套四张关于中东景观的描述已发表在作者 2012 年出版的手绘表达专著中。此后作者随着多次游历欧洲及全国各地，形成对某一地区特定景观的整体印象，这其中最直接的方式是利用飞机到达某一地点降落前的一段时间进行俯拍和整体感受。相对于时间要求严格的起飞阶段，降落具有更长的观察时间和平衡广阔的视野。所以对于观察大地的旅行者有一个忠告，即最好选择飞机后排座位避开所有的安全通道和最主要的障碍物（机翼），同时能够获得比飞机前部更大的视野。

  本章主要选择了江南、荷兰、托斯卡纳、普罗旺斯等世界各地典型的农田景观，讲述了应地形排水，森林－土地轮作等自然和人的因素所造成的独特而富有魅力的景观样式，这是我们当代进行大地艺术探索的最好的老师和最感人的范本。

　　In this chapter, with landscapes under common influence of both human and natural forces as the example, a group of sceneries focusing on the artificial nature, or agricultural landscape are presented. Such kind of sceneries is the most authentic and oldest manifestation of "poetic residence". The creation was inspired by the planning for Jerusalem, Israel, in 2008, in which the author got involved. A set of four sketches about the Middle East landscape was included in the author's hand-painted monograph published in 2012. Afterward, the author has traveled extensively across Europe and across China, and got a whole picture of particular landscape in a region, directly by high angle shot and holistic perception before landing of plane. Virtually, take-off period has very strict time limit, while landing period renders a longer observation time and a broad vision of balance. So there is a tip for the travelers intending to observe the earth, that is, better choose the rear seats to avoid safe passage and main obstacle (wing). In this manner, one may have a broader view than sitting in the front seats.

　　There are mainly typical farmland landscapes in the South China, the Netherlands, Tuscany, Provence and other places around the world, showing the unique, charming landscape styles resulted from various natural and human factors, such as topography-suitable drainage, forest-land crop rotation, etc. These are the best teachers and most impressive model for our contemporary exploration in land art.

左图上下四张讲述了一处天然的地形，在海风和人的共同作用下所经历的变化。由上至下，第一幅为原始的地貌，一处由海风作用影响的植物分布景观，植物大多聚集于受盐碱和毛细作用影响较小的山脊和土层较厚的区域；第二幅在此基础上继续繁衍，可以在一部分汇水区形成更为茂盛的植物群落，大地因植物的更替演化第一次发生变化；第三幅是在人类的影响下，通过全面改造而形成的类似草原风光的景观，这一时期海风和盐碱的影响为人类的改造力所抵消，但是植物的茂密程度，林带的分布远没有达到英国风景园牧场风光所具有的如画境界，可以称为贫瘠的英国牧场风光；第四幅，同样的地块如果出现在美国的南方，如佛罗里达地区，通过高超的营造手段，人类在这一地区营造了数量惊人的、生长良好的高尔夫球道。球道两侧长草区域的大型热带植物体现了最典型的美国南方景观特征，这是英国多云的天空下无法呈现的景观。

The 4 drawings on this page, from top to down, depict the changes of a natural landform under common effects of sea breezes and human activities. From up to down the original landform in different stages are shown. The first picture is plant distribution landscape under influence of sea breezes, where most plants gather on the mountain ridge less affected by saline-alkali and the capillary force, or in zones with relatively thicker soil; the second picture shows the result of reproduction on the basis of picture 1, where more luxuriant plant communities have come into being in some catchment areas, and the ground experienced the first transformation along with the alternation of plants; the third picture shows a landscape that is similar to grassland scenery, which is made through overall transformation by human force; in this stage the effects of sea breezes and saline-alkali are offset by human activities, but the vegetation and forest belt distribution is not as dense as those in the picturesque pasture sceneries of British gardens; so it can be called "barren British pasture scenery"; in the last picture, the landscape of the same plot in somewhere in the South U.S., in Florida for example, is shown, where superior construction methods are utilized by human to build a huge number of well-grown golf courses. Large tropical plants in the light roughs on both sides of the golf courses embody the typical characteristics of the landscape in the South of the U.S., which is a mission impossible under the cloudy sky in Britain.

川西平原浅丘地区特有的农田景观，没有云阳梯田的戏剧化和织锦缎一般的光彩，但温润的气候条件下，长势喜人的梯田依旧体现出人对自然改造的伟大力量。

右图为荷兰填海造田的典型样式，林、田、水依据格栅的网络和人的意志进行最理性化的分布，其中的开阔、壮丽的景观迥然不同于山水相依的中国传统农田景观。

This is the farmland landscape peculiar to shallow hills on the Western Sichuan plain. It is not as dramatic as Yunyang terraced fields nor radiating the brocade-like brilliance, but in the warm climate crops in the terrace fields are growing vibrantly, repaying the great power of human transformation over the nature.

The lower drawing shows the typical style of "reclaiming areas from the sea to create farmlands" in the Netherlands. Forests, fields and waters are arranged in the most rational way by the grille grids and human wills. The broad, magnificent landscape is totally different from the traditional Chinese agricultural landscape where mountains and rivers support each other.

同样类型的浅丘与平原结合的景观，在意大利中部则表现为迥然不同的情景，这里所有向阳的山坡均被经济型的林木所占据，一片片灰蒙蒙的郁郁葱葱的橄榄林构成了地中海天空下最出色的景观群落。每一棵橄榄树、无花果都体现了意大利人兼顾景观的利益与美感的睿智，而这里景观的骨架顺理成章地由苏格兰雪松改造成了意大利特有的柏树。

下图是遍及中南欧几乎所有高速公路两旁的农场景观，大片的棕色的林木和绿色的草场构成了欧洲梦幻农田的骨架，而这每一棵植物几乎都有它的经济利益。地中海灿烂的阳光为这样的农庄吹进了最可贵的生命与鲜活。

The landscape in central Italy, though also a combination of shallow hills and plain, express entirely different scenes, where all sunny mountain slopes are occupied by economic forest trees, and patches of lush, grey olive groves make up the best sight under the sky of Mediterranean. Each olive tree and fig tree reflects the wisdom of Italians in reconciling landscape benefits and aesthetics. Here the cypress peculiar to Italian naturally become the framework of landscape instead of Scots cedar.

The sketch below shows the farm landscape prevailing on almost every highway across Central and Southern Europe. Vast brown trees and green pastures form the skeleton of European fantasy farmland, and almost every plant bears its own economic interest. Blessed by the Mediterranean sunshine, farms get enlivened and prosperous.

杏花春雨江南，所有的一切都被蒙上一层雨雾，天空、水田与浅丘如一幕幕戏曲交错出现，自上而下，体现出极强的形式化特征。

"Apricot blossom with spring rain in South China". As it goes, the world is covered with rain mist. Sky, paddy fields and shallow hills appear alternatively like scenes of a drama, from top to down, reflecting strong formal characteristics.

下图为典型的英国园林风光，是来普顿在红皮书当中所描述的最典型的景观，其中虽然在细节上有浅丘与平原的差别，但是整体风貌依旧，作为景观框架的苏格兰红松，永远都是景观中的守望者和真正的骨干，布朗利用了这种树的神圣的力量配合英国特有的光线，共同塑造了如丝绸一般的浅丘与阴影。可以说这是景观人文主义者心目中英国牧场情结的真正的依托。

Below is a typical British garden scene, the most representative landscape described by Clapton in the Red Book. Some details are different between shallow hills and plains, but not affecting the overall style and features. Scots pine, as the landscape framework, is always the watcher and real backbone in the landscape. By combining the sacred power of Scots pine and the Britain-specific lights, Brown successfully expresses the silk-like shallow hills and shadows. In some sense this really embodies the British pasture complex of landscape humanists.

　　本页图为佛罗伦萨的郊外,以农场、种植园和经营性木草场为景观斑块,数以千计的农场单元,共同构筑了全托斯卡纳令人难忘的农业景观。当你把镜头拉近,就可以从任何一片锦缎一般的土地上清晰地分辨出酿酒的葡萄园,经济型的橄榄园和无花果园各自的特征,比如秋天葡萄园呈现一片金色,橄榄园保持一贯的青灰色,牧草转为一种淡淡的黄绿色,其背后是千家万户的庭园草坪,为大量世界级的酒庄及大量的欧洲农业品牌所共同依托,而他们所组成的综合体又为世界各地的旅行者所赞叹。事实上,这种延绵上千年的意大利农业传统,究其思想深度和影响的广度都远远超过了英国单一的牧场景观的范畴。就绘画的形式角度来说,一个最根本的就是大片延展的农庄与所有道路上笔直竖向天空的所有的柏树的对比,构成了此类景观的共同的框架结构。

　　The drawings on this page and the opposite show the suburb of Florence, with farms, plantations and for-profit pastures as the landscape patches. Thousands of farm units together create the unforgettable agricultural landscape of Tuscany. Zooming in, you can, on any brocade-like land, clearly distinguish from the respective characteristics of wine vineyard, economic olive groves and fig orchards. For example, in the autumn, vineyards become golden, olive groves are livid as they are all the time, and the pasture become light yellow-green, as the background of garden lawns of millions of households. They are the common support for a large number of world class wineries and European agricultural brands. These elements, in turn, compose a complex that is admired by travelers around the world. In fact, in terms of both depth of thoughts and sphere of influence, that agricultural tradition in Italy lasting over thousands of years has far exceeded the scope of single pasture landscape in Britain. From the drawing's formal perspective, the fundamental element is contrast between the vast extending farms and cypress trees on the road straightly stretching toward the sky. This constitutes the common framework for such type of landscape.

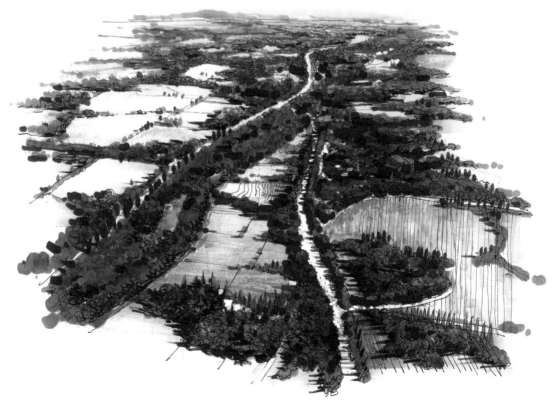

　　黔中之山，山山相连，自安顺到六盘水沿途多喀斯特山形，自空中鸟瞰形如"一锅馒头"，然细致近看，则山形面面有色，面面有形。沿线的郎岱、镇宁等古镇及六枝特区，多位于深山大谷之中，由一水贯穿全境，城郭两侧皆山，山坡中段皆层层梯田穿行其中，如在云端。

In the middle of Guizhou province, there are mountains connected with each other. Along the way from Anshun to Liupanshui many mountains are in the Karst landform, appearing the shape of "a potful of steamed buns" viewed from the sky. However when you observe them closely, you will find the colors and shapes of each facade. Langdai and Zhenning ancient towns, as well as Liuzhi special district along this way are mostly hidden in the mountains and large valleys. A river flows throughout the zone, and on both sides of the city are hills, with tiers of terraced fields on hillsides in the mid-course, creating an atmosphere as in the cloud.

场地现状景观
Current landscape on the venue

合成以后的场地设计概念草图
Composite design conceptual sketch of the venue

场地周边盆景似的卡斯特山地实景
Imaging figure of the bonsai-like Karst topography around the venue

一张普通的现场云海照片
A common photo of the cloud sea at the venue

在 Ps 中合成一张最简易场景图纸
Compose a simplest scene drawing by Photoshop

加画前景
Add the foreground

选取一张合适的远景山脉作为背景
Select a suitable picture of distant mountain range as the background

将农田前景与调整后的远景结合
Combine the farmland foreground with the adjusted distant scenery

选择性地取舍细节，上色并加上设计元素
Add or delete details as needed, and paint color on the picture with design elements

中东意象
Middle East imagery

"一花一世界，一叶一菩提"，这句东方式的禅语用在这几座以色列园林中同样显得十分贴切。历史花园中粗糙的石灰岩墙面，随山势婉转的水渠，规则种植的橄榄、柑橘、无花果等当地最常见的材料和形式，通过符号化的提取叠加，便被赋予了纪念性意义。正是对平凡习见的一砂一石的精致处理，使场地的历史和未来得以联系，也由此塑造出属于耶路撒冷这座伊斯兰教、基督教和犹太教共同的文化中心应有的景观。

The Buddhist allegorical words, "One flower is a world, each leaf has a life", rightly fit these Israeli gardens. The most common local materials, such as rough limestone walls in historic gardens, water channels winding along mountain, regularly planted olives, citrus and figs, carry out commemorative significance accordingly, by virtue of symbolic extraction and superposition. It is the delicate efforts put into those ordinary stone and sand that link history and future of the sites together, and that create an appropriate landscape unique to Jerusalem, the common cultural center of Judaism, Christianity and Islam.

天堂乐园——以色列耶路撒冷城市景观设计国际竞赛

以色列耶路撒冷城市景观设计属于"耶路撒冷城市之门"城市设计国际咨询的第二部分。项目所涉六块主要场地总面积约380hm²，位于耶路撒冷新城南部。

本设计依据现状地形的总体趋势，在380hm²的巨大场地上，由北至南布局了体育公园、博物馆园、历史文化园、高速公路园、中国园和森林公园等六个主体场地，分别容纳了城市休闲、运动，文化交流、教育等功能。项目设计于2008年12月中旬完成，2009年入围国际竞赛前五名，并获荣誉奖。

Paradise City—— International Urban Landscape Design Competition in Jerusalem, Israel

The urban landscape design of Jerusalem, Israel, falls into the part II of the "Jerusalem Gate" International Consultation on the Urban Design. With a total area of about 380 hectares, the six main sites of the Project are located in the south of New City of Jerusalem.

In the design, following the existing overall trend of the landform of the huge field up to 380ha., six main venues are arranged from north to south, namely the Sports Park, the Museum Garden, the Historical and Cultural Garden, the Expressway Garden, the Chinese Garden and the Forest Garden, offering respective functions like urban recreation, sporting, cultural exchange, education and alike. The design was completed in mid-December 2008, and was listed top five for the international competition in 2009, and eventually won the Honorary Award.

设计思想——将场地作为以色列传统农业文化和地理版图的双重表征

（1）充分尊重原有地形，并赋予场地地貌以独特的文化内涵，以景观的手法表达出以色列国土和地形特征。

（2）植物元素的文化意义：植物设计以以色列特有的石松、柏树、橄榄树、柑橘、棕榈、葡萄、薰衣草、无花果为主导，突出以色列悠久的农业文明和发达的现代农业科技。

Design concepts – defining the venue as a dual representation of Israel for both traditional farming culture and the geographical layout

(1) With fully respect for the original landform, endow the venue's landscape unique cultural connotations, and express the features of Israel's territory and landform by way of landscape.

(2) The cultural meaning of plant elements: the plant design is dominated by species peculiar to Israel, such as club moss, cypress trees, olive trees, citrus trees, palms, grapes, lavender and figs, so as to highlight the long-lasting agricultural civilizations of Israel and its developed modern agricultural science and technologies.

体育公园

此公园保留现有的体育设施和活动场地,在现状植物稀疏的场地增加了足球场、野餐区,为周边高度密集的社区居民提供了更多的户外空间。在服务设施方面,设置了充足的绿荫停车场地和必要的设施,如小型购物场所、小型餐馆、厕所、自行车租赁等。在景观方面,结合现状地形及现状植物,增加了一些趣味的景观元素,如迷宫花园,更开辟了一些林间空间为人们野餐及自由活动提供了场所。

1. Existing Sports C
2. Parking Lot with Brick Pavement
3. Flower Shrubs
4. Service Facilitie
5. Camp fire
6. Maze Garden
7. Toilet
8. Restaurant

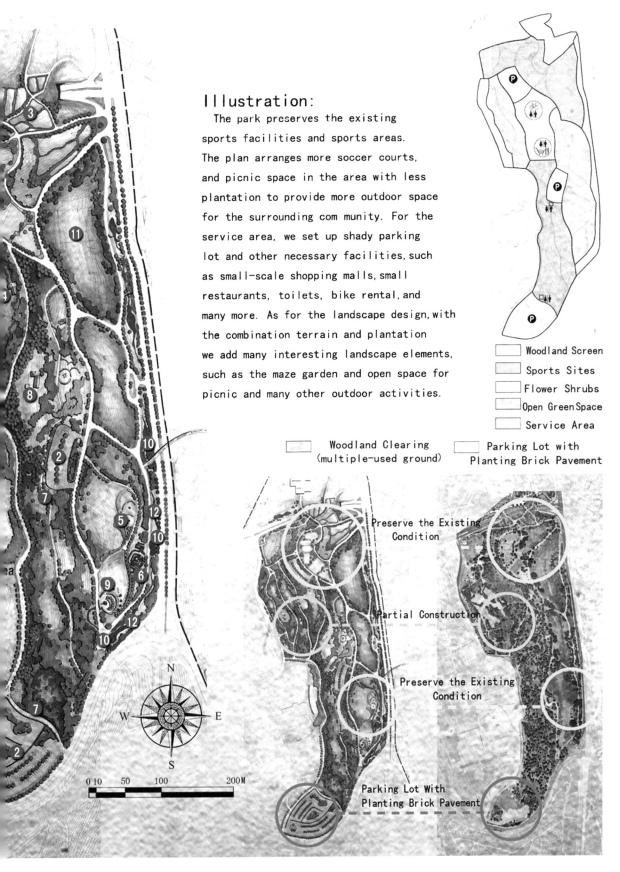

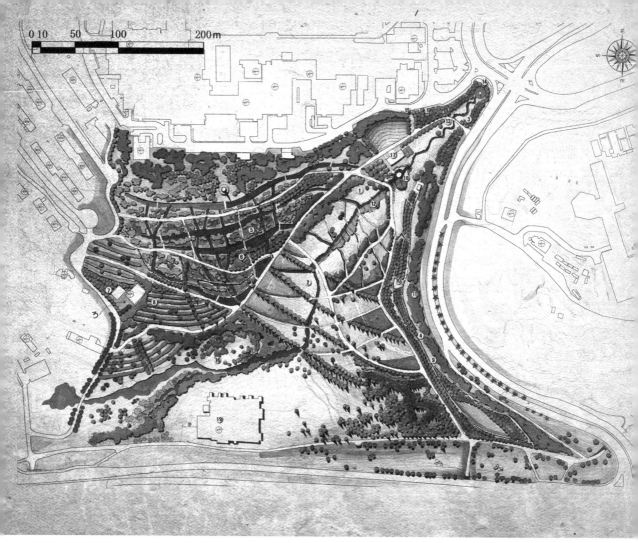

历史文化园平面图
Plan of the Historical and Cultural Park

历史文化园

　　历史文化园植物设计的灵感来自于圣经中提到的，被认为是富有以色列地域性的七种水果和谷物，包括小麦、大麦、葡萄、无花果、石榴、橄榄和椰枣。除此七种植物以外，在保留现有植被和地形的基础上，还加入了其他一些当地树种，象征了以色列的农业、历史和文化传统。

　　花园水渠设计充分结合了雨水管理和灌溉功能，线性水系沿地形汇集成小溪流，一直延伸到教堂前的水花园，此水花园为农神萨顿之所在；另一支水渠由东北角的涌泉汇集而来，沿线状地形和一系列小瀑布叠水缓缓流入水花园中。

　　用水渠划分场地、联络空间的做法充分考虑了场地的历史感和文化属性，也考虑了耶路撒冷这座对于犹太教、伊斯兰教和基督教均具有极端重要意义的城市所应有的景观形式和能反映场地历史文脉延续性的文化符号。与东方的中国园林和日本园林所追求的不规则布局、自由形态相反，伊斯兰园林和基督教文化下的园林基于一种精心设计的直线型和四分园的基本构成模式：由中心引出的四条水渠伸向花园四方，将一块场地划分为四个方形园林（或苗圃），以此象征天国世界的基本模式。这四条水渠在印度莫卧儿王朝的伊斯兰园林中，干脆被称为"水、酒、乳、蜜"四河，而四条"河流"交汇之处必然会有甘冽的清泉喷涌而出，这便是人类的天堂。

The Historical and Cultural Park

The design was inspired by seven fruits and grains mentioned in the Bible and are believed to be representative in Israel, including wheat, barley, grape, fig, pomegranate, olive and date palm. In addition to these plants and preserved existing vegetation and landforms, a number of other local species are also included to symbolize the agricultural, historical and cultural traditions of Israel.

Garden channels fully combine rainwater management and irrigation systems. With such design, the linear water system forms a small stream along the terrain and extends to the water garden, which is in front of church and houses Saturn — god of agriculture and vegetation. Another channel from the bubbling spring at the northeastern corner slowly flows into the water garden along the linear terrain and a series of cascading waterfalls.

The design of utilizing channels to divide sites and link spaces takes full account of the historic and cultural attributes of sites, the proper landscape form fitting the holy city in the three major Abrahamic religions of Judaism, Christianity and Islam, and cultural symbols that can reflect continuity of the site's historical context. Quite an opposite to the irregular layout and free form pursued by oriental Chinese gardens and Japanese gardens, the Islamic gardens and Christian gardens are based on a well-designed linear and Char Bagh Garden composition: four central channels stretch toward four sides of gardens to divide a site into four square gardens (or nurseries), symbolizing the basic model of the divine world. In Islamic gardens of Mughal Empire in India, these four channels are simply called "water, wine, milk, honey" rivers, at the intersection of which there must emerge clear spring — a paradise of mankind.

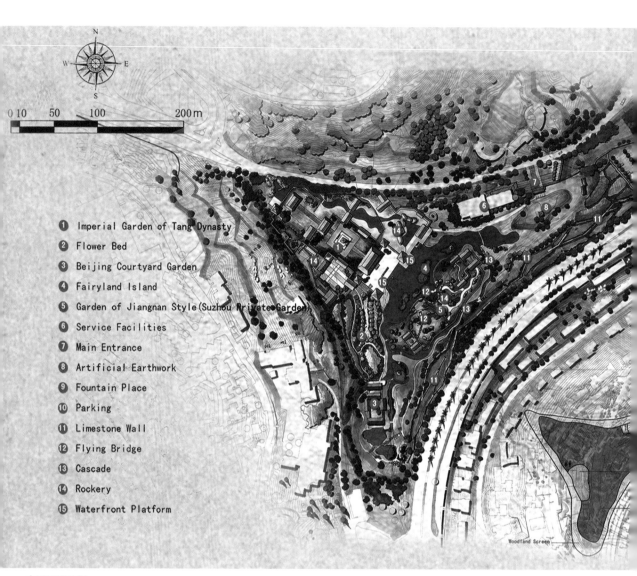

中国园平面图
Plan of the Chinese Garden

① Imperial Garden of Tang Dynasty
② Flower Bed
③ Beijing Courtyard Garden
④ Fairyland Island
⑤ Garden of Jiangnan Style (Suzhou Private Garden)
⑥ Service Facilities
⑦ Main Entrance
⑧ Artificial Earthwork
⑨ Fountain Place
⑩ Parking
⑪ Limestone Wall
⑫ Flying Bridge
⑬ Cascade
⑭ Rockery
⑮ Waterfront Platform

中国园

在一片四周高中间低洼的三角地块上设计一个中心汇水区域，成为中国园的主湖，并使之成为园中所有建筑和假山叠石的共同背景。围绕主湖和由之向西、南伸展出的二条溪流、叠水串联了三组庭院和建筑，即大唐中国园、苏州园和一个典型的北京四合院庭院，分别体现轴向、非轴向自由空间和围墙为特色的容积空间，三种典型的中国空间形式。

三组空间成为展示多种历史悠久的中国工艺、文明、民俗的舞台，如中国雕刻、剪纸、泥塑、叠石、书、画、墨、砚等具有中国园林特色的植物，如黑松、白皮松、侧柏、牡丹、竹、梅和最重要的石组等，将成为景观空间最主要的表情，将原植物园的植物基底延伸入三座庭院外围，形成新旧景观之间的联系和桥梁。

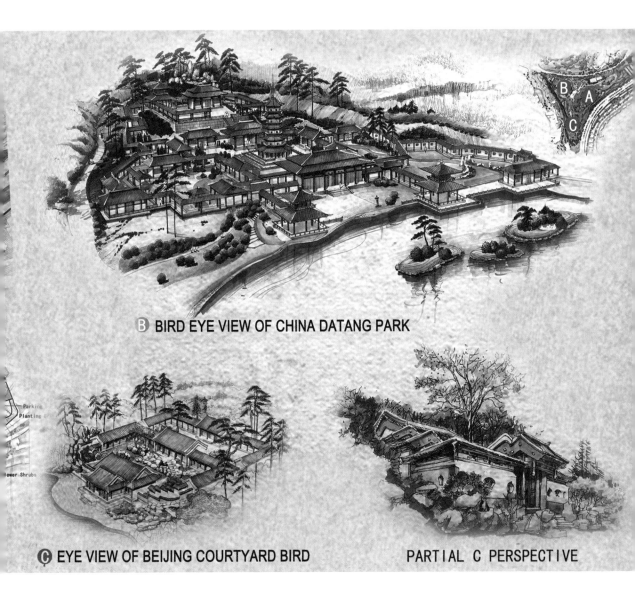

B BIRD EYE VIEW OF CHINA DATANG PARK

C EYE VIEW OF BEIJING COURTYARD BIRD

PARTIAL C PERSPECTIVE

The Chinese Garden

The site is a triangle-shaped plot with a low-lying central part and higher surrounding; so the center is designed into a catchment area, which is the main lake of the garden, as well as the background shared by all architectures, rockeries and balanced rocks in the garden. Two streams runs around the main lake and stretch toward the west and south respectively. The connect three sets of courtyards and architectures, namely, the China Datang Garden, the Suzhou Garden and a typical Beijing Siheyuan courtyard. They respectively reflect three typical Chinese spacial forms, the axial and non-axial free spaces, as well as volume spaces featuring enclosing walls.

Three sets of spaces serve as a stage displaying time-honored Chinese technology, civilization and folklore, such as Chinese carving, paper cutting, clay sculpture, stacked stone, calligraphy, painting, ink and inkstone. Becoming a dominant expression of landscape space, plants with Chinese garden characteristics, such as black pine, white bark pine, platycladus orientalis, peony, bamboo, plum and the most important stone group, are to extend the plant base of the original botanical garden into surroundings of three courtyards, thus acting as a bridge between the old and new landscapes.

南宁 2018 国际园林博览会大师园规划鸟瞰（设计者，中国工程院院士孟兆祯）
Aerial view of the Master Garden in 2018 China (Nanning) International Garden Expo, designed by Meng Zhaozhen, (An) academician of China National Engineering Research Institute

西藏林芝鲁朗林海度假小镇规划鸟瞰
Aerial view of the Holiday town planning of Lin lulang, Nyingchi, Tibet

图书在版编目（CIP）数据

主题场景/王劲韬著. —北京：中国建筑工业出版社，2017.9
景观设计手绘教程
ISBN 978-7-112-21008-4

Ⅰ.①主… Ⅱ.①王… Ⅲ.①景观设计－绘画技法－教材 Ⅳ.① TU986.2

中国版本图书馆CIP数据核字（2017）第176749号

责任编辑：杜　洁　段　宁　李玲洁
责任校对：焦　乐

景观设计手绘教程

主题场景

王劲韬　著

*

中国建筑工业出版社出版、发行（北京海淀三里河路9号）
各地新华书店、建筑书店经销
北京方舟正佳图文设计有限公司制版
深圳市泰和精品印刷厂印刷

*

开本：787×1092毫米　1/16　印张：8½　字数：207千字
2017年9月第一版　2017年9月第一次印刷
定价：68.00元
ISBN 978-7-112-21008-4
（30629）

版权所有　翻印必究
如有印装质量问题，可寄本社退换
（邮政编码 100037）